Wallace

First Published 2023
© text Aneirin Karadog, 2023
© illustrations Alyn Smith, 2023

No part of this publication may be reproduced, stored in a retrieval system, or transmitted, in any form, or by any means, electrical, mechanical, photocopying, recording or otherwise without the prior permission of the publisher or a licence permitting restricted copying.

ISBN 978-1-91-430328-9

Published by Llyfrau Broga Books, Whitchurch, Cardiff

www.broga.cymru

Wallace

The Curious Life of Alfred Russel Wallace

Written by Aneirin Karadog
Illustrated by Alyn Smith

Alfred Russel Wallace was born in Usk in 1823, one of nine children.

Because the Wallace family didn't have much money, Alfred attended school for only six years. He spent lots of his time outdoors or studying old books and maps.

When he was fourteen, Alfred went to live in London with his brother John, a trainee carpenter. He loved to learn and read all the books he could find – he even sneaked into places to hear talks about science, nature and society.

It was a time when scientists and explorers were heroes.

Alfred dreamed of one day joining the Royal Society, a group of the most important scientists in the world – people like Charles Darwin.

Alfred moved back to Wales to work as a map-maker.

For eight years he walked for miles around the countryside; all the time he was measuring, drawing and writing notes.

Alfred loved nature and used his wanderings to study the wildlife wherever he went.

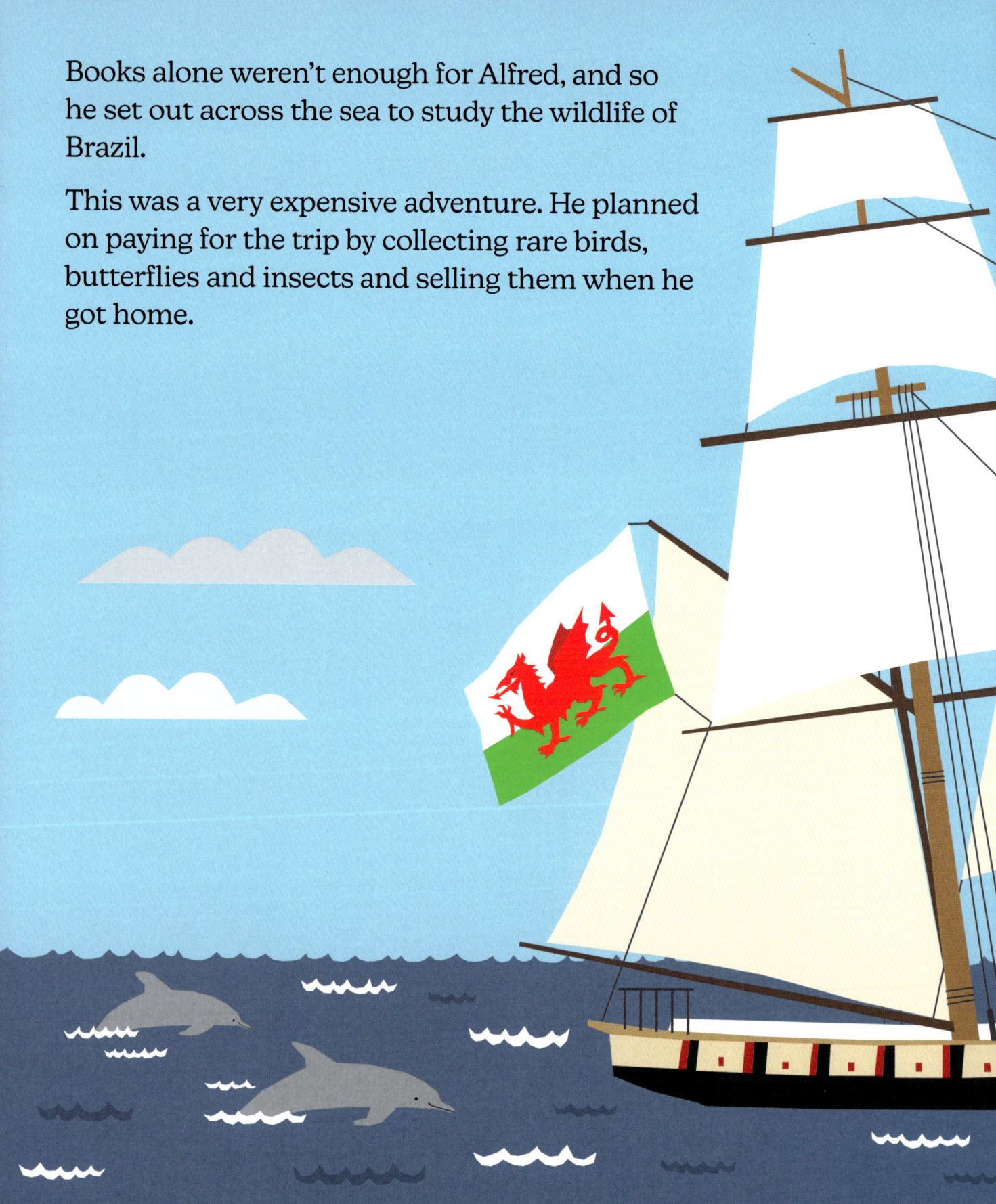

Books alone weren't enough for Alfred, and so he set out across the sea to study the wildlife of Brazil.

This was a very expensive adventure. He planned on paying for the trip by collecting rare birds, butterflies and insects and selling them when he got home.

The main reason for the journey was to try and find the answer to a problem that puzzled scientists: where had all the different types of animals and plants come from?

This was called The Problem of the Origin of Species.

Many other explorers of the time thought they were better than the native people they met. Alfred was different. He studied their languages and habits and could see that they were just as clever and able as he was.

After four years mapping, drawing and collecting specimens deep in the Amazon rainforest, where no other scientist had ever been, Alfred was ready to go home.

On the journey back, disaster struck. A fire started and everyone had to abandon ship. All of the animals and insects, maps and notes that Alfred had collected were lost in the fire.

The crew spent ten days on the open sea in a lifeboat before being rescued.

Alfred did manage to save part of his diary and a few sketches.

From these he wrote two books about the Amazon, which helped to pay for his next trip – to the Malay islands.

He still hadn't lost his love of adventure!

Alfred spent eight years in the Malay islands, finding thousands of new types of birds and animals.

He also drew maps and searched for clues to solve the mystery of where all the different species came from.

One day, when ill in bed with a tropical fever, Alfred finally found his answer; new species appear as animals slowly change over a long time to have a better chance of survival in the place where they live.

Alfred sent his answer to his hero, Charles Darwin, and the two scientists together wrote the first book explaining the idea – which became known as 'evolution by natural selection'.

It was an idea that changed the world.

Alfred returned to Britain for good in 1862. He married and had three children.

Their house had a wonderful garden – and space to hold his amazing collection of over 125,000 specimens.

Alfred spoke out on many different topics: the environment, alien life and even spirits and ghosts.

When he was seventy, and one of the world's experts on animal and plant life, Alfred was finally awarded a place in the Royal Society.

Alfred Russel Wallace was one of the most important scientists of his time.

He always thought for himself rather than following what everyone else believed, even though this sometimes made him unpopular.

Alfred's life was built around his love of wildlife and care for people ... and his endless curiosity.

Read about more
Welsh Wonders

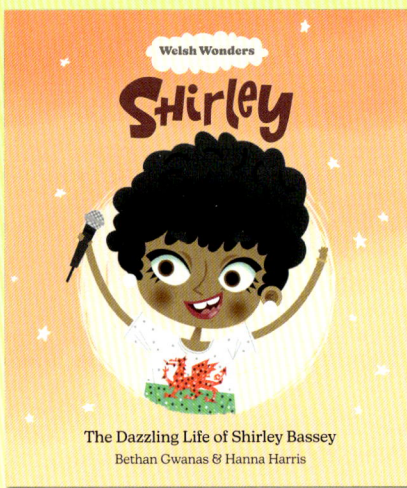

Shirley Bassey
The girl from Tiger Bay whose voice became famous around the world.

Cranogwen
Sarah Jane Rees was a sea captain, prize-winning poet, publisher, and inspiration!

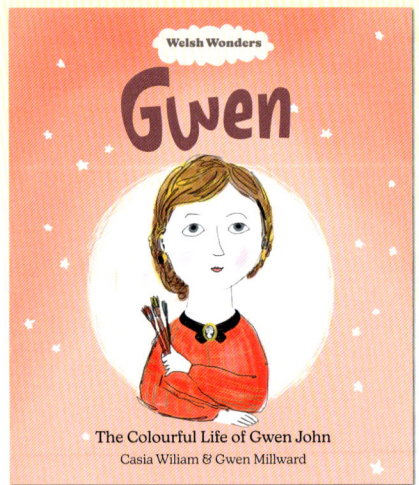

Gwen John
A shy but determined girl who loved to paint and followed her dream of being a famous artist.

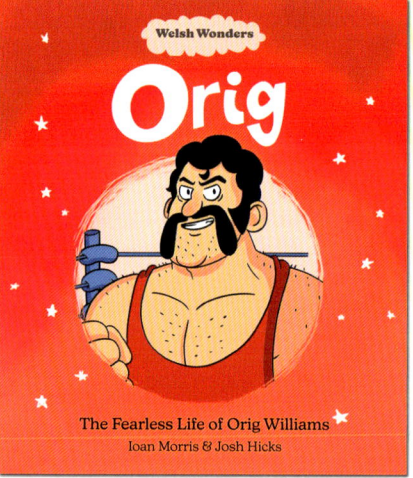

Orig Williams
The tough-guy wrestler with a heart of gold, known around the world as El Bandito!

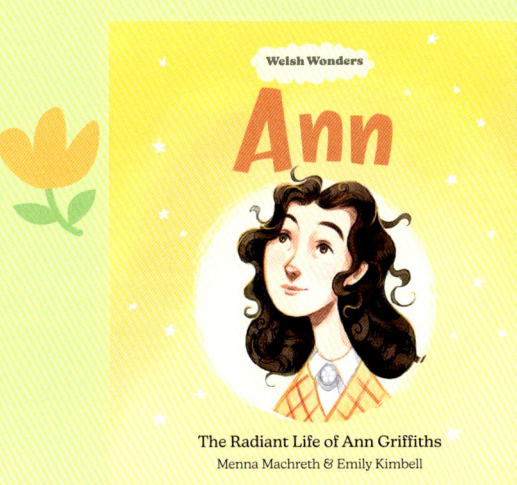

Ann Griffiths
The sensitive poet whose spiritual songs inspired millions.

Aneurin Bevan
Inspirational politician who founded the NHS and changed a nation.

Laura Ashley
Fashion designer who built a business empire from her home in mid Wales.

Betty Campbell
The inspirational story of Wales' first Black headteacher, who fought for equality and fairness in education.

Find out more about other inspiring Welsh lives – from artists and scientists to people who challenged the way things were and overcame difficulties to achieve their dreams.